Ministry of Agriculture, Fisheries and Food
Agricultural Development and Advisory Service
Land and Water Service

GW00372447

The design of field drainage pipe systems

Reference Book **345**

London: Her Majesty's Stationery Office

© *Crown copyright 1982*
First Published 1982

ISBN 0 11 241515 6

Contents

Introduction

This publication provides the essential design elements for determining the pipe sizes required for field drainage systems, whether the pipes are used as laterals, mains, interceptors, piped ditches or culverts.

The design methods are approximate and the procedures are as simple as possible and compatible with convenient and speedy day-to-day use. The methods can be used for the generality of problems likely to be met, but more complex problems of an exceptional nature will require special consideration falling outside the scope of this document.

The range of charts provided covers most types of pipe commonly in current use.

If additional information is required regarding the detailed basis from which the charts and appendices have been derived, enquiries should be addressed to the appropriate ADAS drainage specialist at any MAFF divisional office.

1.0 Pipe size for a lateral

1.1 Definition

A lateral is a sub-surface pipeline which collects soil water continuously over its length, either through gaps at pipe joints or slots in the pipe wall.

1.2 Information Needed

1.2.1 Intended land use, ie grassland, arable horticultural crop etc.
1.2.2 Soil type, ie likely rate of water movement – hydraulic conductivity.
1.2.3 Rainfall probability. See MAFF Technical Bulletin No 34, Climate and Drainage (Reference 1), or the simplified extracts at Appendix 1 of this note.
1.2.4 The required pipe gradient.

1.3 Procedure

1.3.1 From Appendix 1, determine the design rainfall (r), in mm/day appropriate to:
    The district concerned
    Intended type of land use
    Type of drainage treatment
1.3.2 From Appendix 2, determine the drainflow factor (f) taking into account:
    The future land use
    Type of drainage system
    Subsoil permeability class
    Land surface slope

1.3.4 From Appendix 3, determine the catchment per lateral (A_L), in ha from drain spacing and maximum required length of lateral.
1.3.5 Using the results from Appendices 1, 2 and 3 above, compute the design flow (Q_L) in litre/s from:
$$Q_L = 0.13 \times r \times f \times A_L$$
1.3.6 Decide on the required pipe gradient (%).
1.3.7 Apply Q_L and pipe gradient to Charts 1-4 according to the type of pipe to be used, to determine the required pipe size. If the intersection on the chart lies between pipe sizes select the larger of the two sizes.

2.0 Pipe size for a restricted inlet main

2.1 Definition

A restricted inlet main is a sub-surface pipeline which intercepts laterals and conveys the combined flow to an outfall. The main may have gaps or slots in it and thereby itself act as a collector of soil water throughout its length, as do laterals.

2.2 Information Needed

2.2.1 Intended land use, ie grass, horticulture etc.
2.2.2 Soil type, ie likely rate of water movement – hydraulic conductivity.
2.2.3 Rainfall probability. See MAFF Technical Bulletin No 34, *Climate and Drainage* (Reference 1), or the simplified extracts at Appendix 1 of this note.
2.2.4 The required pipe gradient.
2.2.5 The total catchment area measured to the proposed pipeline outfall.

2.3 Procedure

2.3.1 From Appendix 1, determine the design rainfall (r), in mm/day, appropriate to:
    The district concerned
    Intended type of land use
    Type of drainage treatment
2.3.2 From Appendix 2, determine the drainflow factor (f) taking into account:
    The future land use
    Type of drainage system
    Subsoil permeability class
    Land surface slope

2.3.3 Determine the total catchment area in ha (A_M) to be drained by the pipeline in question, measured to its outfall.

2.3.4 Using the above results compute design flow (Q_M) in litre/s from:

$$Q_M = 0.13 \times r \times f \times A_M$$

2.3.5 Decide on the required pipe gradient (%).

2.3.6 Apply Q_M and pipe gradient to Charts 5-10 according to the type of pipe to be used, to determine the required pipe size. If the intersection on the chart lies between pipe sizes select the larger of the two sizes. Where the design velocity exceeds 1.5 m/s pipe joints should be sealed.

3.0 Pipe size for an interceptor drain

3.1 Definition

An interceptor drain is a layout of one or more pipe drains located to intercept areas of seepage from springs or springlines.

3.2 Procedure

The theoretical design of interceptor drains is inherently complex. The relevant catchment area is difficult or impossible to determine with any degree of accuracy and the relationship between design flows and rainfall is virtually indeterminate. The selection of pipe size should therefore always err on the side of being generous. In the absence of other information the design should be based on the data in Appendix 4 and the pipe design Charts 1-4 as appropriate.

4.0 Pipe size for an open-inlet piped ditch

4.1 Definition

An open-inlet piped ditch is the replacement of an open ditch with a line of pipes, or otherwise a line of pipes which receives a ditch discharge at its inlet and conveys water to a point of outfall.

4.2 Scope

The method described is only suitable for catchments not exceeding 30 ha.

4.3 Informaion Needed

Catchment characteristics, ie
    Catchment area
    Maximum length of catchment
    Average slope
    Dominant crop type
    Soil type
    Percentage paved areas in the catchment
    Average annual rainfall

4.4 Procedure

4.4.1 Using Appendices 5 and 6, determine the design flow Q_O in litre/s.

4.4.2 Decide on the required pipe gradient (%).

4.4.3 Apply Q_O and pipe gradient to Charts 5, 8, 9 or 10 according to the type of pipe to be used, to determine the required pipe size. If the intersection on the chart lies between pipe sizes select the larger of the two sizes.

5.0 Pipe size for a culvert

5.1 Definition

A pipe culvert is the piping of a short length of ditch in order to provide access over it.

5.2 Procedure

For the small catchments normally associated with field drainage culverts the problem can be regarded as a short length of open-inlet piped ditch and the design procedure outlined in paragraph 4.4 above adopted accordingly.

6.0 References

1 MAFF Technical Bulletin No 34, *Climate and Drainage*, HMSO 1976.

Appendix 1.1

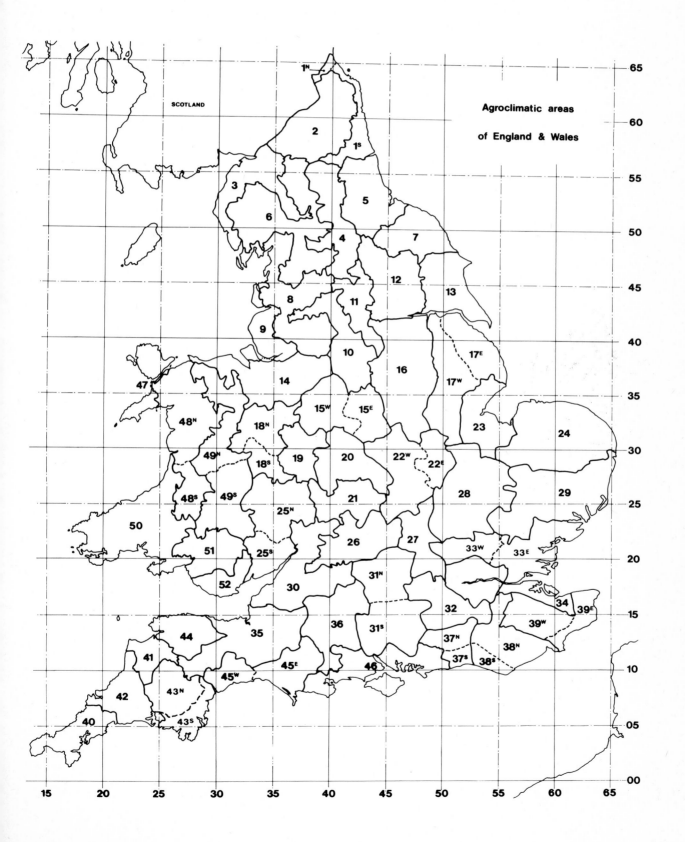

Agroclimatic areas
of England & Wales

SCOTLAND

Appendix 1.2

Daily Rainfall for Field Drainage Design (mm/day)

Agro-climatic Area No	Mean Annual Rainfall (mm)	Mole drainage schemes			Other types of drainage		
		Grassland	*Arable*	*Horticulture*	*Grassland*	*Arable*	*Horticulture*
1 N	651	17	21	28	6	8	11
1 S	670	18	22	29	7	8	11
2	939	27	35	45	12	14	17
3	1045	28	35	45	12	14	18
4	1068	31	38	50	13	16	21
5	669	18	22	31	7	9	12
6	1663	40	48	64	17	20	25
7	808	23	27	36	9	11	15
8	1133	32	37	47	13	15	18
9	837	23	27	33	9	11	13
10	1151	36	41	52	15	17	21
11	807	25	31	40	10	12	15
12	643	19	23	31	7	9	11
13	655	18	21	27	7	8	10
14	786	20	25	34	8	10	12
15W	840	25	28	28	9	11	14
15E	714	20	25	34	8	9	12
16	622	18	23	32	7	8	11
17W	605	17	22	31	7	8	10
17E	648	18	22	31	7	8	11
18N	763	21	26	33	9	10	12
18S	845	24	29	37	10	11	11
19	700	18	22	32	7	9	11
20	695	19	23	31	8	9	11
21	660	18	22	31	7	9	11
22W	661	19	22	30	7	8	10
22E	627	18	21	29	7	8	9
23	575	16	20	27	6	7	9
24	623	16	20	28	6	7	10
25N	746	22	27	34	10	11	14
25S	953	29	34	43	13	15	18
26	726	21	25	33	8	10	13

Appendix 1.3

Daily Rainfall for Field Drainage Design (mm/day)

Agro-climatic Area No	Mean Annual Rainfall (mm)	Mole drainage schemes			Other types of drainage		
		Grassland	Arable	Horticulture	Grassland	Arable	Horticulture
27	669	19	22	30	7	9	11
28	574	15	19	26	6	7	9
29	598	16	19	28	6	7	10
30	775	22	27	35	9	10	13
31N	677	19	23	30	8	9	12
31S	798	25	31	37	10	12	15
32	713	20	24	33	8	10	13
33E	577	15	19	26	6	7	9
33W	664	18	21	29	7	8	11
34	630	18	22	28	7	9	11
35	865	24	29	38	11	12	15
36	799	23	27	35	10	11	14
37N	778	24	29	37	10	12	15
37S	820	26	31	39	11	13	16
38N	791	24	30	34	11	13	15
38S	836	25	31	37	12	14	16
39W	696	21	26	34	9	10	13
39E	683	20	25	33	8	10	13
40	1047	30	34	44	13	14	17
41	1133	35	41	54	17	18	22
42	1241	35	40	52	15	18	22
43N	1449	44	50	65	21	24	28
43S	1048	34	39	49	15	17	21
44	1265	36	45	60	18	20	24
45W	910	29	33	43	13	15	17
45E	885	29	34	44	13	14	17
46	807	25	29	36	11	13	16
IOW	794	25	30	38	10	13	16
47	1004	26	30	40	11	13	16
48N	1829	45	50	61	19	22	26
48S	1632	40	45	56	17	19	24
49N	1184	30	35	45	13	15	18
49S	1190	30	35	45	13	15	18
50	1258	32	36	47	13	15	19
51	1729	42	48	58	18	21	25
52	1172	30	34	44	13	15	18

Appendix 2

Drainflow factors (f)

Drainage System	Land Use/ Cultivation	Slope <1%			1%-3%			>3%		
		Subsoil Permeability Class								
		Rapid	Med	Slow	Rapid	Med	Slow	Rapid	Med	Slow
Pipes only	Arable	1.0	0.8	0.7	0.9	0.7	0.6	0.7	0.6	0.5
	Grass	0.9	0.7	0.6	0.8	0.6	0.5	0.6	0.5	0.4
Pipes* and subsoiling	Arable	—	0.9	0.8	—	0.8	0.7	—	0.7	0.6
	Grass	—	0.8	0.7	—	0.7	0.6	—	0.6	0.5
Pipes and Moling	Arable	—	—	0.8	—	—	0.8	—	—	0.7
	Grass									

*This means true subsoiling where the intention is to shatter the soil. If a subsoiler tine is used to create a discrete channel ie 'square moling' – the design system should be 'Pipes and Moling'.

Appendix 3

Ready reckoner for lateral length, drain spacing and area; maximum permissible lateral length in metres

Area (Ha)	Drain spacing (m)							
	40	30	20	18	16	14	12	10
1.6	400							
1.5	375							
1.4	350							
1.3	325							
1.2	300	400						
1.1	275	367						
1.0	250	333						
0.95	238	317						
0.90	225	300						
0.85	213	283						
0.80	200	267	400					
0.75	188	250	375					
0.70	175	233	350	389				
0.65	163	217	325	361				
0.60	150	200	300	333	375			
0.55	138	183	275	306	344	393		
0.50	125	167	250	278	313	357		
0.49	123	163	245	272	306	350		
0.48	120	160	240	267	300	343	400	
0.47	118	157	235	261	294	336	392	
0.46	115	153	230	256	288	329	383	
0.45	113	150	225	250	281	321	375	
0.44	110	147	220	244	275	314	367	
0.43	108	143	215	239	269	307	358	
0.42	105	140	210	233	263	300	350	
0.41	103	137	205	228	256	293	342	
0.40	100	133	200	222	250	286	333	400
0.39	98	130	195	217	244	279	325	390
0.38	98	127	190	211	238	271	317	380
0.37	93	123	185	206	231	264	308	370
0.36	90	120	180	200	225	257	300	360
0.35	88	117	175	194	219	250	292	350
0.34	85	113	170	189	213	243	283	340
0.33	83	110	165	183	206	236	275	330
0.32	80	107	160	178	200	229	267	320
0.31	78	103	155	172	194	221	258	310
0.30	75	100	150	167	188	214	250	300
0.29	73	97	145	161	181	207	242	290
0.28	70	93	140	156	175	200	233	280
0.27	68	90	135	150	169	193	225	270
0.26	65	87	130	144	163	186	217	260
0.25	63	83	125	139	156	179	208	250
0.24	60	80	120	133	150	171	200	240
0.23	58	77	115	128	144	164	192	230
0.22	55	73	110	122	138	157	183	220
0.21	53	70	105	117	131	150	175	210
0.20	50	67	100	111	125	143	167	200
0.19	48	63	95	106	119	136	158	190
0.18	45	60	90	100	113	129	150	180
0.17	43	57	85	94	106	121	142	170
0.16	40	53	80	89	100	114	133	160
0.15	38	50	75	83	94	107	125	150
0.14	35	47	70	78	88	100	117	140
0.13	33	43	65	72	81	93	108	130
0.12	30	40	60	67	75	86	100	120
0.11	28	37	55	61	69	79	92	110
0.10	25	33	50	56	63	71	83	100
0.09	23	30	45	50	56	64	75	90
0.08	20	27	40	44	50	57	67	80
0.07	18	23	35	39	44	50	58	70
0.06	15	20	30	33	38	43	50	60
0.05	13	17	25	28	31	36	42	50

Appendix 4

Inflow rate per 100m of drain (litre/s)

Soil Texture	Slope			
	<2%	2-5%	5-12%	>12%
Coarse sand and gravel	1.3-9.0	1.4-10.0	1.6-11.0	1.7-12.0
Sandy loam	0.6-2.2	0.7- 2.4	0.7- 2.6	0.8- 2.9
Silt loam	0.4-0.9	0.4- 1.0	0.4- 1.1	0.5- 1.2
Clay and clay loam	0.2-0.6	0.2- 0.7	0.2- 0.7	0.2- 0.8

Appendix 5

Determination of design flow

1. Locate a suitable map of the area and determine the catchment area A in hectares.

2. Determine the maximum length of catchment L in metres

3. Determine the average slope of the catchment S

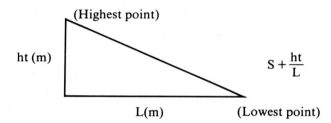

(Highest point)

ht (m)

L(m)   (Lowest point)

$$S + \frac{ht}{L}$$

4. Determine the catchment of characteristc C

$$C = 0.0001\frac{L}{S}$$

5. Determine the dominant crop type   Grass
                                           Arable
                                           Horticulture

6. Determine the average annual rainfall AAR in mm from Appendix 1

7. Determine the soil type factor S_T

Permeability Class	Range (m/day)	Ranges for Soil Textures	S_T
—	—	Peat (Upland)	1.3
Very slow	<0.01- 0.1	C	1.0
Slow - Mod	0.1 - 0.3	ZyC	0.8
Moderate	0.3 - 1.0	CL SC	0.5
Mod - Rapid	1.0 -10.0	Pt S SL ZyCL	
Very Rapid	>10	(low)	0.1

Note: for complex areas interpolation between soil types may be desirable

8. At Appendix 6 enter the graph at C. Move across (left) to crop type, down to average annual rainfall (AAR), across (right) to the standard line and up to F number.

9. Peak Flood Flow $Q_O = S_T \times F \times A$

10. For each per cent of paved areas add 1% to the derived Peak Flood Flow. (Where paved area exceeds 10% of the catchment this method is not appropriate).

A(ha)	
L(m)	
S	
C(m)	
	G A H
AAR (mm)	
S_T	
F	
Q_O (l/s)	
Total Q_O (l/s)	

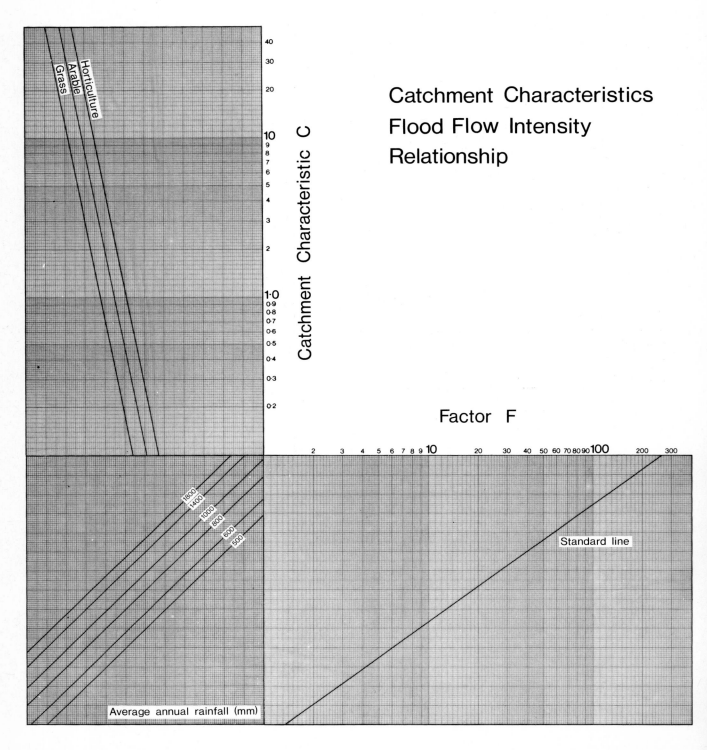

Catchment Characteristics
Flood Flow Intensity
Relationship

Chart 1

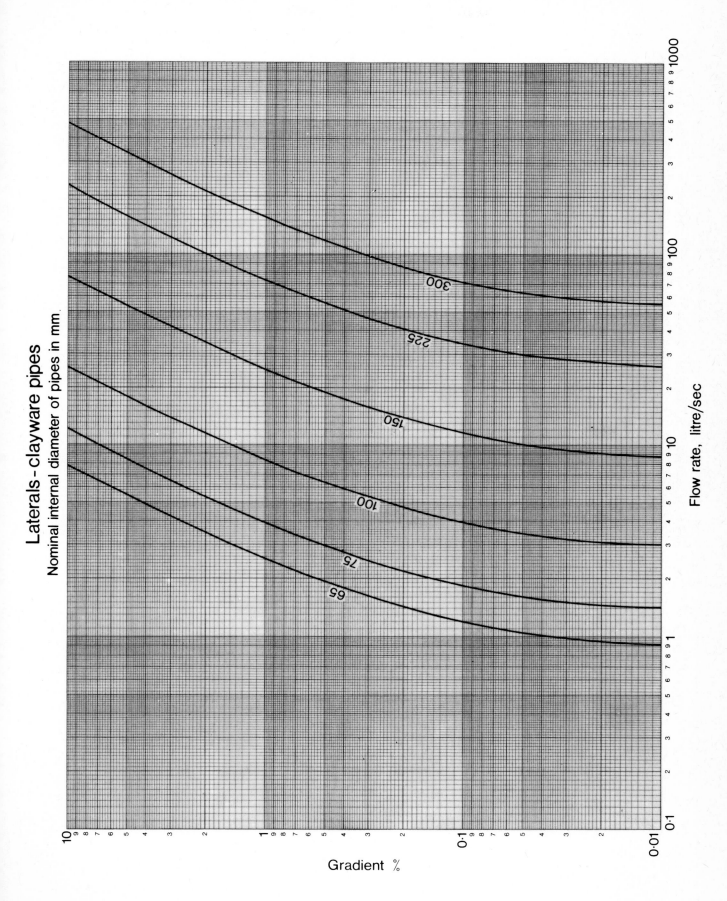

Laterals – clayware pipes
Nominal internal diameter of pipes in mm.

Gradient %

Flow rate, litre/sec

11

Chart 2

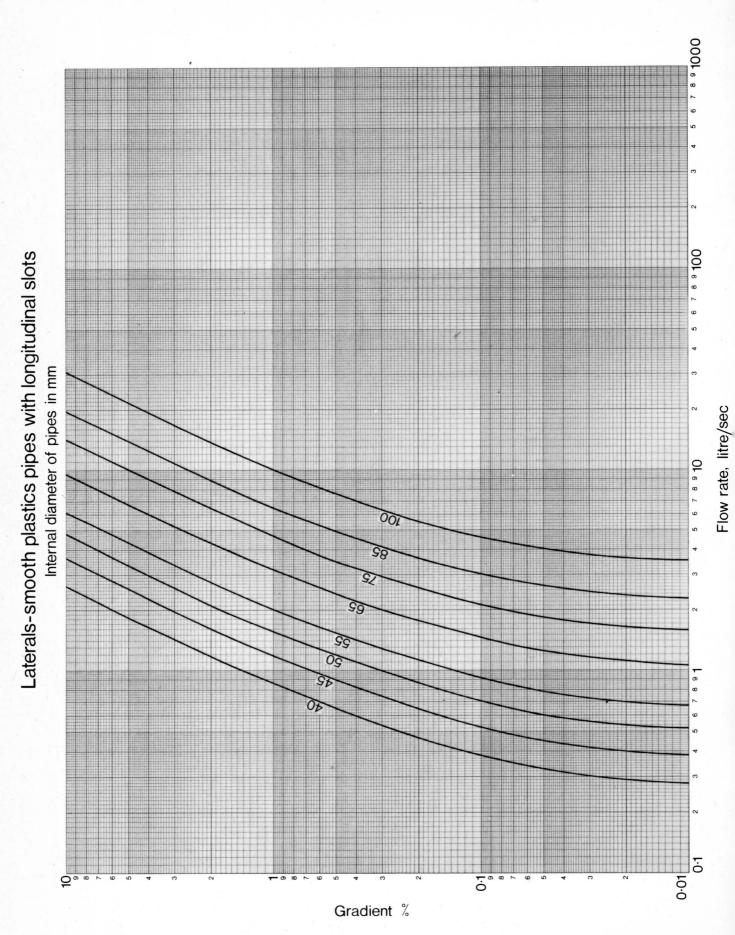

Laterals-smooth plastics pipes with longitudinal slots

Internal diameter of pipes in mm

Flow rate, litre/sec

Gradient %

12

Chart 3

Laterals - smooth plastics pipes with transverse slots
Internal diameter of pipes in mm

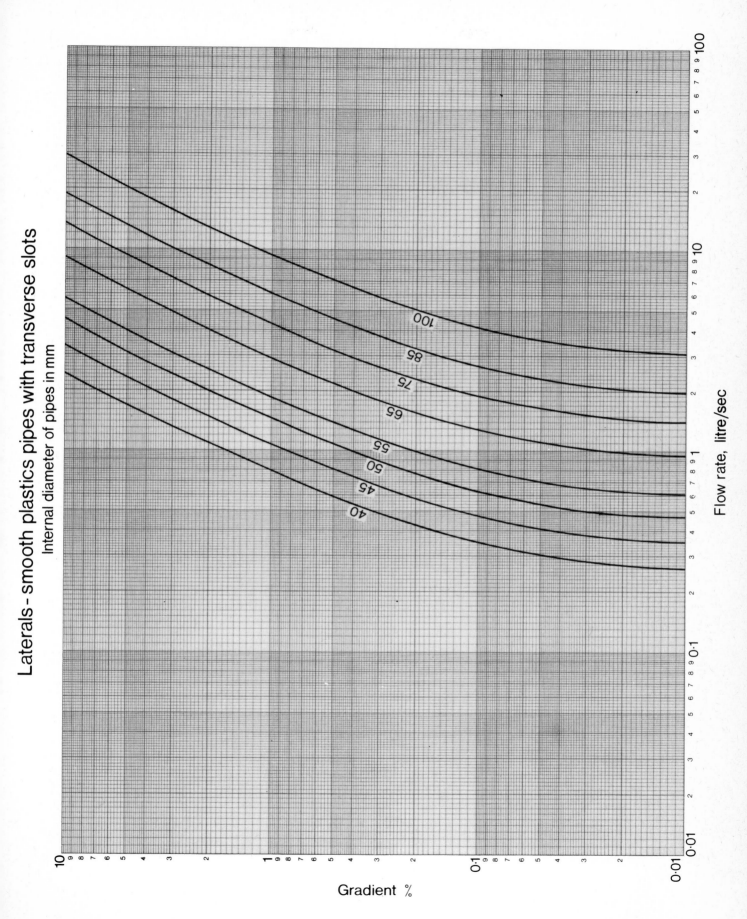

Flow rate, litre/sec

Gradient %

Chart 4

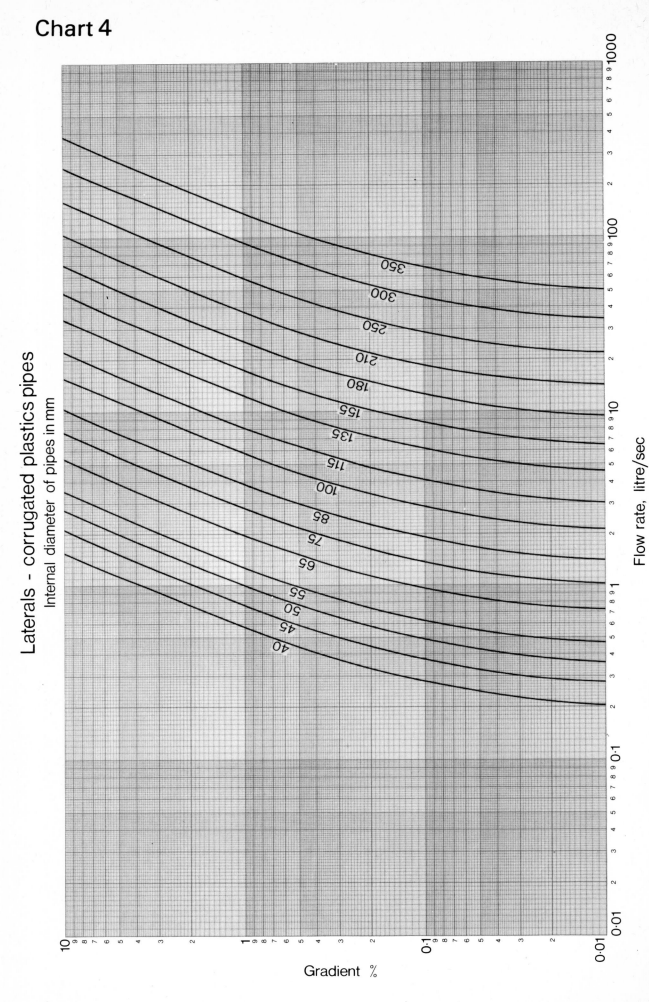

Laterals - corrugated plastics pipes
Internal diameter of pipes in mm

Flow rate, litre/sec

Gradient %

Chart 5

Restricted and open inlet-clayware pipes
Nominal internal diameter of pipes in mm

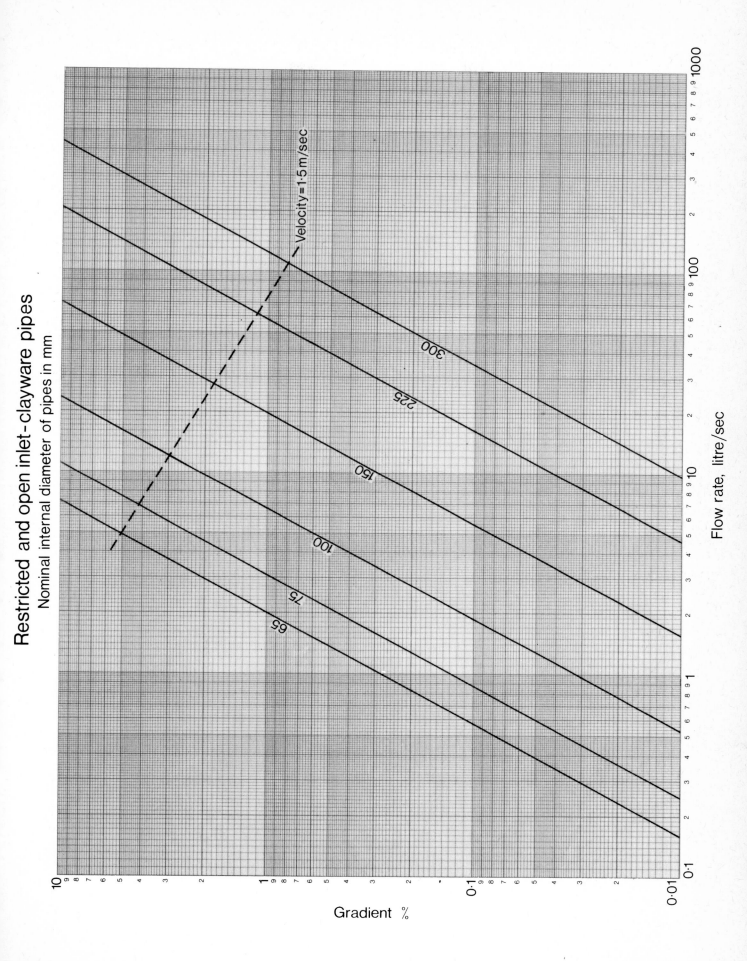

Velocity=1·5 m/sec

300
225
150
100
75
65

Gradient %

Flow rate, litre/sec

Chart 6

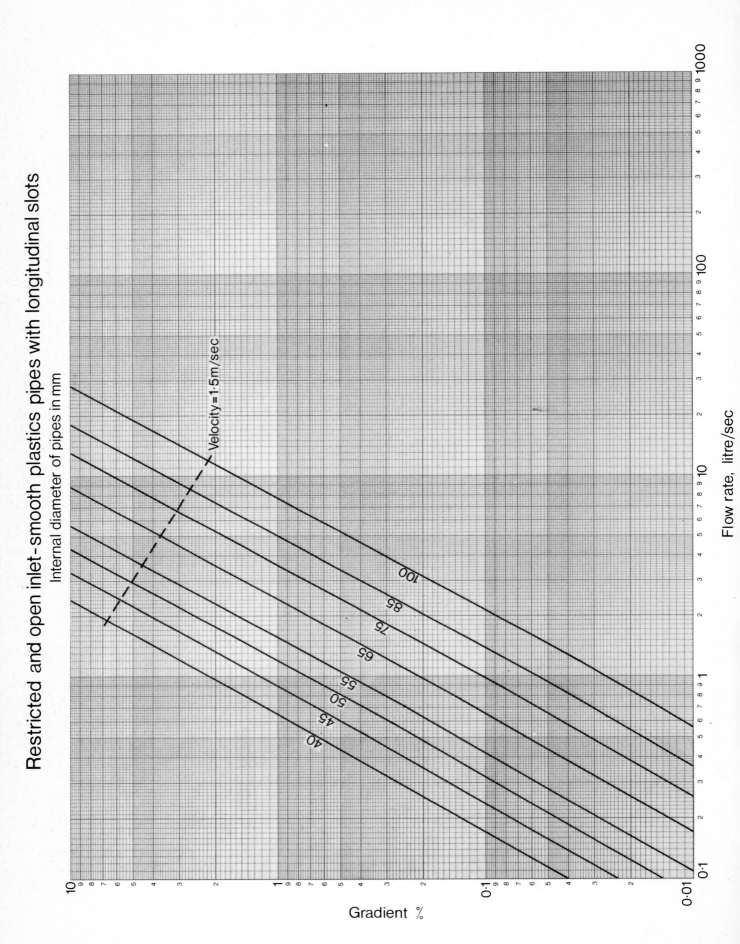

Restricted and open inlet-smooth plastics pipes with longitudinal slots

Internal diameter of pipes in mm

Velocity = 1·5 m/sec

Flow rate, litre/sec

Gradient %

16

Chart 7

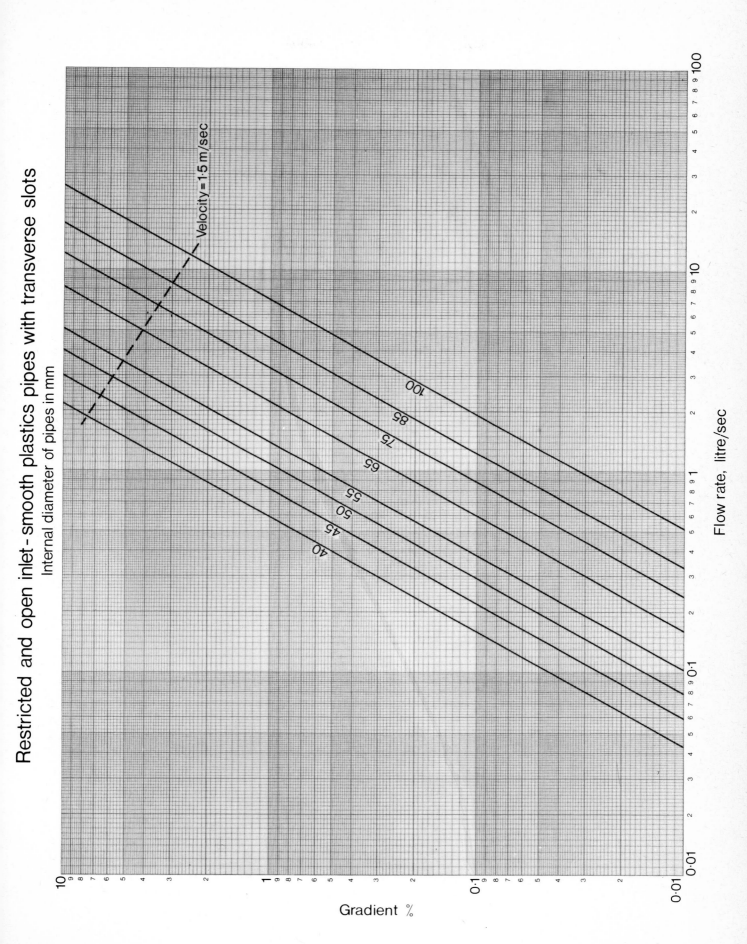

Restricted and open inlet - smooth plastics pipes with transverse slots

Internal diameter of pipes in mm

Velocity = 1·5 m/sec

100
85
75
65
55
50
45
40

Gradient %

Flow rate, litre/sec

17

Chart 8

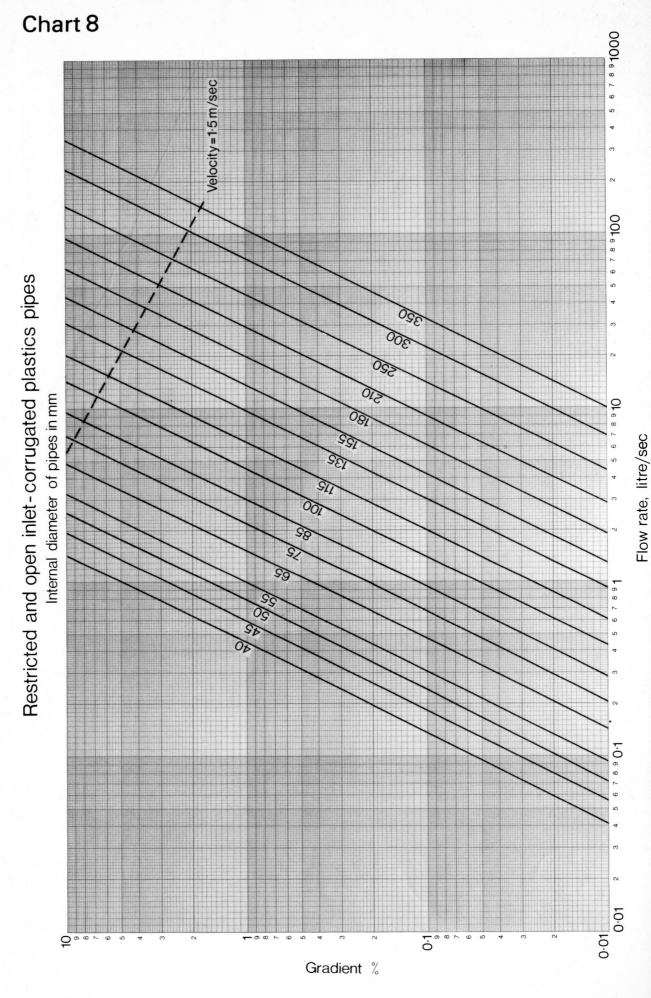

Restricted and open inlet - corrugated plastics pipes
Internal diameter of pipes in mm

Velocity = 1·5 m/sec

Gradient %

Flow rate, litre/sec

350
300
250
210
180
155
135
115
100
85
75
65
55
50
45
40

Chart 9

Restricted and open inlet-smooth uncorrugated pipes
e.g. glazed vitrified clay, plastics and pitch fibre

Internal diameter of pipes in mm

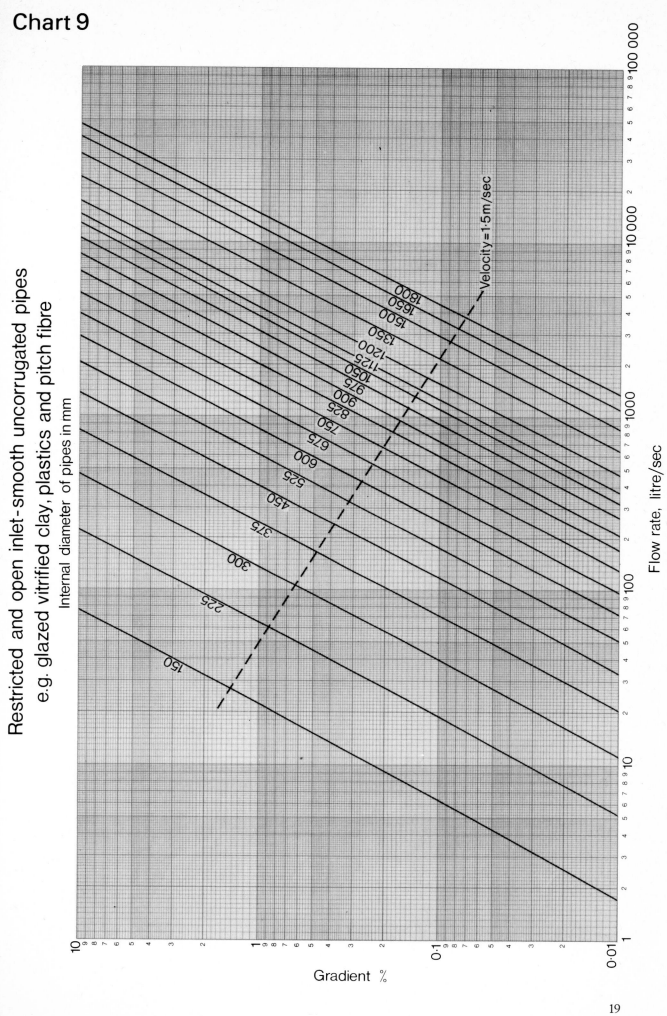

Chart 10

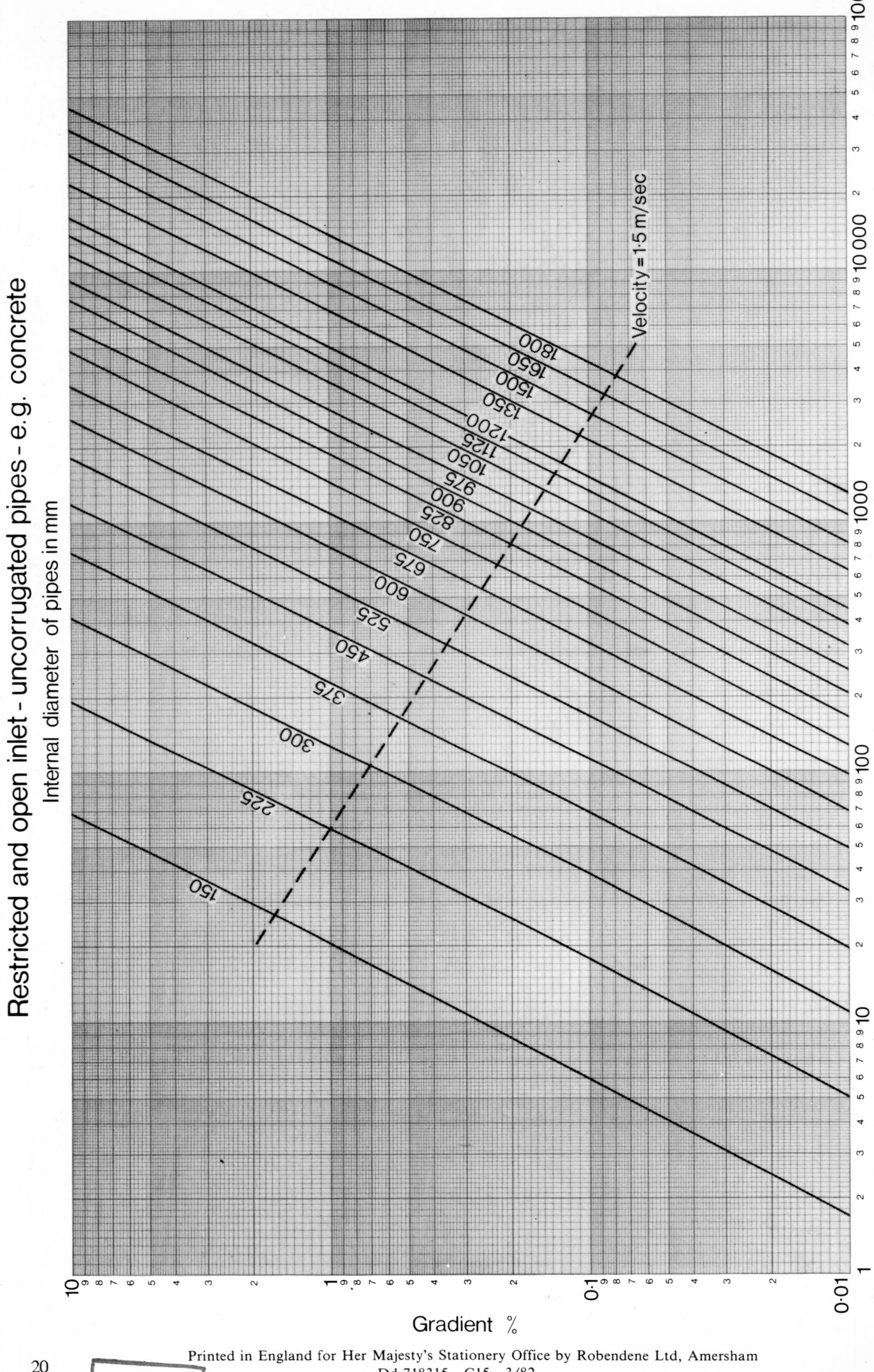

Restricted and open inlet - uncorrugated pipes - e.g. concrete
Internal diameter of pipes in mm
Velocity = 1·5 m/sec
1800
1650
1500
1350
1200
1125
1050
975
900
825
750
675
600
525
450
375
300
225
150
Gradient %
10
1
0·1
0·01
Flow rate, litre/sec
1
10
100
1000
10 000
100 000

Printed in England for Her Majesty's Stationery Office by Robendene Ltd, Amersham
Dd 718315 C15 3/82